Deriving Laws from Ordering Relations

Kevin H. Knuth

Deriving Laws from Ordering Relations

Kevin H. Knuth

Computational Sci. Div., NASA Ames Research Ctr., M/S 269-3, Moffett Field CA 94035

Abstract. The effect of Richard T. Cox's contribution to probability theory was to generalize Boolean implication among logical statements to degrees of implication, which are manipulated using rules derived from consistency with Boolean algebra. These rules are known as the sum rule, the product rule and Bayes' Theorem, and the measure resulting from this generalization is probability. In this paper, I will describe how Cox's technique can be further generalized to include other algebras and hence other problems in science and mathematics. The result is a methodology that can be used to generalize an algebra to a calculus by relying on consistency with order theory to derive the laws of the calculus. My goals are to clear up the mysteries as to why the same basic structure found in probability theory appears in other contexts, to better understand the foundations of probability theory, and to extend these ideas to other areas by developing new mathematics and new physics. The relevance of this methodology will be demonstrated using examples from probability theory, number theory, geometry, information theory, and quantum mechanics.

INTRODUCTION

The algebra of logical statements is well-known and is called *Boolean algebra* [1, 2]. There are three operations in this algebra: conjunction $\wedge$, disjunction $\vee$, and complementation $\sim$. In terms of the English language, the logical operation of conjunction is implemented by the grammatical conjunction '*and*', the logical operation of disjunction is implemented by the grammatical conjunction '*or*', and the logical complement is denoted by the adverb '*not*'. Implication among assertions is defined so that a logical statement a implies a logical statement b, written $a \to b$, when $a \vee b = b$ or equivalently when $a \wedge b = a$. These are the basic ideas behind Boolean logic.

The effect of Richard T. Cox's contribution to probability theory [3, 4] was to generalize Boolean implication among logical statements to degrees of implication represented by real numbers. These real numbers, which represent the degree to which we believe one logical statement implies another logical statement, are now recognized to be equivalent to probabilities. Cox's methodology centered on deriving the rules to manipulate these numbers. The key idea is that these rules must maintain consistency with the underlying Boolean algebra. Cox showed that the product rule derives from associativity of the conjunction, and that the sum rule derives from the properties of the complement. Commutativity of the logical conjunction leads to the celebrated Bayes' Theorem. This set of rules for manipulating these real numbers is not one of set of many possible rules; it is the *unique generalization consistent with the properties of the Boolean algebraic structure.*

Boole recognized that the algebra of logical statements was the same algebra that described sets [1]. The basic idea is that we can exchange 'set' for 'logical statement', 'set intersection' for 'logical conjunction', 'set union' for 'logical disjunction', 'set complementation' for 'logical complementation', and 'is a subset of' for 'implies' and you have the same algebra. We exchanged quite a few things above and its useful to break them down further. We exchanged the *objects* we are studying: 'sets' for 'logical statements'. Then we exchanged the *operations* we can use to combine them, such as 'set intersection' for 'logical conjunction'. Finally, we exchanged the means by which we *order* these objects: 'is a subset of' for 'implies'. The obvious implication of this is that Cox's results hold equally well for defining measures on sets. That is we can assign real numbers to sets that describe the degree to which one set is a subset of another set. The algebra allows us to have a sum rule, a product rule, and a Bayes' Theorem analog—just like in probability theory!

It has been *recognized* for some time that there exist relationships between other theories and probability theory, but the underlying reasons for these relationships have not been well understood. The most obvious example is quantum mechanics, which has much in common with probability theory. However, it is clearly not probability theory since quantum amplitudes are complex numbers rather than real numbers. Another example is the analogy recognized by Carlos Rodríguez between the cross-ratio in projective geometry and Bayes' Theorem [5]. In this paper, I will describe how Cox's technique can be further generalized to include other algebras and hence other problems in science and mathematics. I hope to clear up some mysteries as to why the same basic structure found in probability theory appears in other contexts. I expect that these ideas can be taken much further, and my hope is that they will eventually lead to new mathematics and new physics. Last, scattered throughout this paper are many observations and connections that I hope will help lead us to different ways of thinking about these topics.

LATTICES AND ALGEBRAS

Basic Ideas

The first step is to generalize the basic idea behind the elements described in our Boolean algebra example: objects, operations, and ordering relations. We start with a set of objects, and we select a way to compare two objects by deciding whether one object is 'greater than' another object. This means of comparison allows us to order the objects. A set of elements together with a *binary ordering relation* is called a *partially ordered set*, or a *poset*. It is called a *partial* order to allow for the possibility that some elements in the set cannot be directly compared. In our example with logical implication, the ordering relation was 'implies', so that if 'a implies b', written $a \rightarrow b$, then b is in some sense 'greater than' a, or equivalently, a is in some sense 'included in' b. An ordering relation is generally written as $a \leq b$, and read as 'b includes a' or 'a is contained in b'. When $a \leq b$, but $a \neq b$ then we write $a < b$, and read it as 'a is properly contained in b'. Last, if $a < b$, but there does not exist any element x in the set P such that $a < x < b$, then we say that 'b covers a', and write $a \prec b$. In this case, b is an immediate successor to a in the hierarchy imposed by the ordering relation. It should be stressed that this notation is

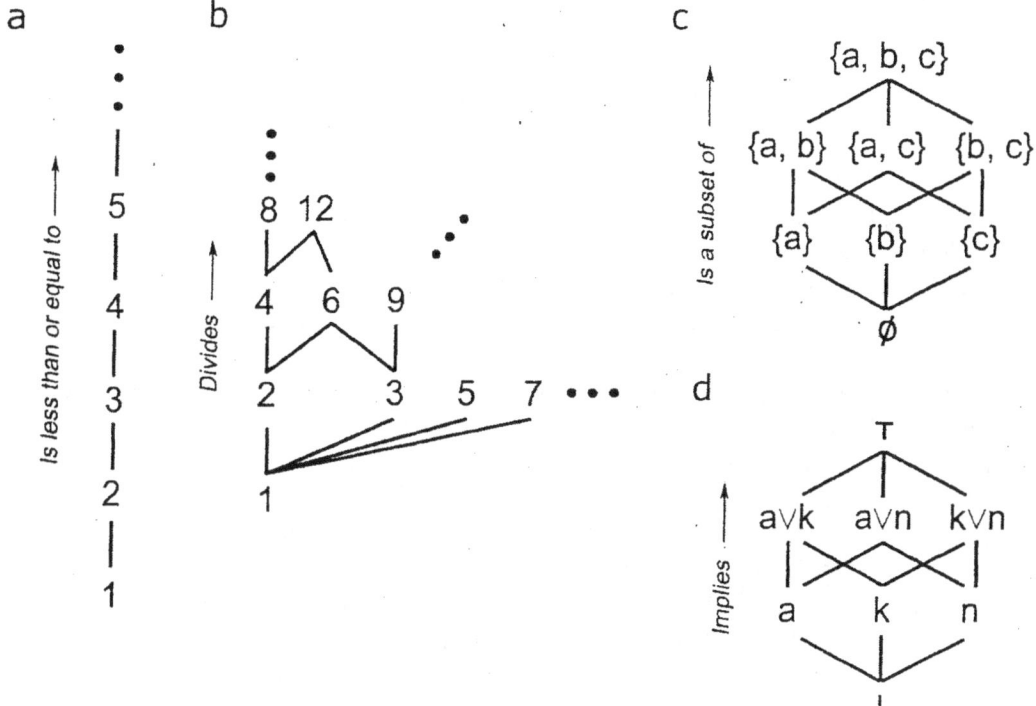

FIGURE 1. Four different posets. (a) The positive integers ordered by 'is less than or equal to'. (b) The positive integers ordered by 'divides'. (c) The powerset of $\{a,b,c\}$ ordered by 'is a subset of'. (d) Three mutually exclusive logical statements a, k, n ordered by 'implies'. Note that the same set can lead to different posets under different ordering relations (a and b), and that different sets under different ordering relations can lead to isomorphic posets (c and d).

general for all posets, but that the relation $\leq$ means different things in different examples.

We can draw diagrams of posets using the ordering relation and the concept of covering. If for two elements a and b, $a \leq b$ then we draw b above a in the diagram. If $a \prec b$ then we connect the elements with a line. Figure 1 shows examples of four simple posets. Figure 1a is the set of positive integers ordered according the the usual relation 'is less than or equal to'. From the diagram one can see that $2 \leq 3$, $2 \prec 3$, but that $2 \not\prec 4$. Figure 1b shows the set of positive integers ordered according to the relation 'divides'. In this poset $2 \leq 8$ means that 2 divides 8. Also, $4 \prec 12$ since 4 divides 12, but there does not exist any positive integer x, where $x \neq 4$ and $x \neq 12$, such that 4 divides x and x divides 12. This poset is clearly important in number theory. Figure 1c shows the set of all subsets (the *powerset*) of $\{a,b,c\}$ ordered by the usual relation 'is a subset of', so that in this case $\leq$ represents $\subseteq$. From the diagram one can see that $\{a\} \subseteq \{a,b,c\}$, and that $\{b\}$ is covered by both $\{a,b\}$ and $\{b,c\}$. There are elements such as $\{a\}$ and $\{b\}$ where $\{a\} \not\subseteq \{b\}$ and $\{b\} \not\subseteq \{a\}$. In other words, the two elements are incomparable with respect to the ordering relation. Last, Figure 1d shows the set of three mutually exclusive assertions a, k, and n ordered by logical implication. These assertions, discussed in greater detail in [6], represent the possible suspects implicated in the theft of the tarts made by the Queen of Hearts in Chapters XI and XII of Lewis

$$a = \text{'Alice stole the tarts!'}$$
$$k = \text{'The Knave of Hearts stole the tarts!'}$$
$$n = \text{'No one stole the tarts!'}$$

Logical disjunctions of these mutually exclusive assertions appear higher in the poset. The element $\bot$ represents the absurdity, and $\top$ represents the disjunction of all three assertions, $\top = a \vee k \vee n$, which is the truism.

There are two obvious ways to combine elements to arrive at new elements. The *join* of two elements a and b, written $a \vee b$, is their least upper bound, which is found by finding both a and b in the poset and following the lines upward until they intersect at a common element. Dually, the *meet* of a and b, written $a \wedge b$, is their greatest lower bound. In Figure 1c, $\{a\} \vee \{b\} = \{a,b\}$, and $\{a,b\} \wedge \{b,c\} = \{b\}$. In that poset, the join $\vee$ is found by the set union $\cup$ and the meet $\wedge$ by set intersection $\cap$. For the poset of logical statements in Figure 1d the notation is a bit more transparent as the join is the logical disjunction (OR). For the meet we have, $(a \vee k) \wedge (k \vee n) = k$, which is the logical conjunction (AND). In Figure 1b, $\wedge$ represents the greatest common divisor $gcd()$; whereas $\vee$ represents the least common multiple $lcm()$. In Figure 1a, $2 \vee 3 = 3$, and $2 \wedge 3 = 2$. In this case $\vee$ acts as the $max()$ function selecting the greatest element; whereas $\wedge$ acts as the $min()$ function selecting the least element. Again, it is important to keep in mind that the symbols $\vee$ and $\wedge$ mean different things in different posets.

Some posets possess a top element, which is called *top* and is generally written as $\top$. The bottom element, called *bottom*, is generally written as $\bot$ or equivalently $\varnothing$. There is an important set of elements in the poset called the *join-irreducible elements*. These are elements that cannot be written as the join of two other elements in the poset. The bottom is never included in the join-irreducible set. In Figure 1c, there are three join-irreducible elements $\{a\}, \{b\}, \{c\}$.

The join-irreducible elements that cover the bottom element are called the *atoms*. In Figure 1b, the atoms are the prime numbers, and the join-irreducible elements are powers of primes. Furthermore, two primes p and q are relatively prime if their meet is the bottom element $p \wedge q = \bot$, that is if $gcd(p,q) = 1$. [1] In Figure 1d, the atoms are the the exhaustive set of mutually exclusive assertions a, k and n. The join-irreducible elements are extremely important as all the elements in the poset can be formed from joins of the join-irreducible elements.

These examples show that the same set under two different ordering relations can result in two different posets (Figures 1a and b), and that two different sets under different ordering relations can result in isomorphic posets (Figures 1c and d). I have discussed these ideas before [6, 8], and one can search out more details in the accessible book by Davey & Priestly [9], and the classic by Birkhoff [10].

---

[1] This is reminiscent of the notation advocated by Graham, Knuth, & and Patashnik [7, p.115] where $p \perp q$ denotes that p and q are relatively prime.

Lattices

Posets have the following properties. For a poset P, and elements $a, b, c \in P$,

P1. *For all a, $a \le a$*   (*Reflexive*)
P2. *If $a \le b$ and $b \le a$, then $a = b$* (*Antisymmetry*)
P3. *If $a \le b$ and $b \le c$, then $a \le c$,* (*Transitivity*)

If a unique meet $x \wedge y$ and join $x \vee y$ exists for all $x, y \in P$, then the poset is called a *lattice*. Each lattice L is actually an *algebra* defined by the operations $\vee$ and $\wedge$ along with any other relations induced by the structure of the lattice. Dually, the operations of the algebra uniquely determine the ordering relation, and hence the lattice structure. Viewed as operations, the join and meet obey the following properties for all $x, y, z \in L$

L1. $x \vee x = x, \quad x \wedge x = x$ (*Idempotency*)
L2. $x \vee y = y \vee x, \quad x \wedge y = y \wedge x$ (*Commutativity*)
L3. $x \vee (y \vee z) = (x \vee y) \vee z, \quad x \wedge (y \wedge z) = (x \wedge y) \wedge z$ (*Associativity*)
L4. $x \vee (x \wedge y) = x \wedge (x \vee y) = x$ (*Absorption*)

There is a special class of lattices called *distributive lattices* that follow

D1. $x \wedge (y \vee z) = (x \wedge y) \vee (z \wedge y)$ (*Distributivity of $\wedge$ over $\vee$*)

and its dual

D2. $x \vee (y \wedge z) = (x \vee y) \wedge (z \vee y)$ (*Distributivity of $\vee$ over $\wedge$*)

All distributive lattices can be expressed in terms of elements consisting of sets where the join $\vee$ and the meet $\wedge$ are identified as the set union $\cup$ and set intersection $\cap$, respectively.

Some distributive lattices possess the property called *complementation* where for every element x in the lattice, there exists a unique element $\sim x$ such that

C1. $x \vee \sim x = \top$
C2. $x \wedge \sim x = \bot$

Boolean lattices are *complemented distributive lattices*. Since all distributive lattices can be described in terms of a lattice of sets, the condition of a distributive lattice being complemented is equivalent to the condition that the lattice contains all possible subsets of the lattice elements, which is called the *powerset*. Thus lattices of powersets are Boolean lattices. The situation gets interesting when one starts removing elements from the powerset. For example, if one element is removed from a Boolean lattice, then there will be another element that no longer has a unique complement. This is equivalent to adding constraints, and these constraints take the Boolean lattice to a distributive lattice, which no longer has complementation as a general property.

These are basic mathematical concepts and are not restricted to the area of logical inference. Viewed as a collection of partially ordered objects, we have a lattice. Viewed as a collection of objects and a set of operations, such as $\vee$ and $\wedge$, we have an algebra. What I will show in the remainder of this paper is that a given lattice (or equivalently its algebra) can be extended to a calculus using the methodology introduced by Cox, and that there already exist a diverse array of examples outside the realm of probability theory.

DEGREES OF INCLUSION

For distinct x and y in a poset where y includes x, written $x \leq y$, it is clear that x does not include y, $y \nleq x$. However, we would like to generalize this notion of inclusion so that even though x does not include y, we can describe the *degree* to which x includes y. This idea is perhaps made more clear by thinking about a concrete problem in probability theory. Say that we know that a logical statement $a \vee b \vee c$ is true. Clearly, $a \rightarrow a \vee b \vee c$ since $a \leq a \vee b \vee c$. However, it is useful to quantify the *degree* to which $a \vee b \vee c$ implies a, or equivalently the degree to which a includes $a \vee b \vee c$. It is in this sense that we aim to generalize inclusion on a poset to degrees of inclusion.

The goal of this paper is to emphasize that inclusion on a poset can be generalized to degrees of inclusion, which results in a set of rules or laws by which these degrees may be manipulated. These rules are derived by requiring consistency with the structure of the poset. At this stage, it is not clear for exactly what types of posets or lattices this can be done, precisely what rules one obtains and under what conditions, and exactly what types of mathematical structures can be used to describe these degrees of inclusion. These remain open questions and will not be explicitly considered in this paper.

What is clear is that Cox's basic methodology of deriving the sum and product rules of probability from consistency requirements with Boolean algebra [3, 4] has a much greater range of applicability than he imagined. His specific results are restricted to complemented lattices as he used the property of complementation to derive the sum rule. In the following sections, I will consider the larger class of distributive lattices, which include Boolean lattices as a special case. To derive the laws governing the manipulation of degrees of inclusion, I will rely on the proofs introduced by Ariel Caticha [11], which utilize consistency with associativity and distributivity. As I will show below, the sum rule is consistent with associativity of the join, and therefore most likely enjoys a much greater range of applicability—perhaps extending to all lattices. The derivations that follow will focus on finite lattices, although extensions to infinite lattices is reasonable.

Joins and the Sum Rule

Throughout this subsection, I will follow and expand on Caticha's derivation and maintain consistency with his basic notation, all the while considering this endeavor as a generalization of inclusion on a poset. We begin by defining a function ϕ that assigns to a pair of elements x and y of the lattice L a real number[2] $d \in \mathbb{R}$, so that $\phi : L \times L \rightarrow \mathbb{R}$

$$d = \phi(x, y), \tag{1}$$

---

[2] For simplicity I will work with degrees of inclusion measured by real numbers. However, it should be kept in mind that Caticha's derivations were developed for complex numbers [11], Aczél's solutions for the associativity equation were for real numbers, groups and semigroups [12], and Rota's theorems for the valuation equation apply to commutative rings with an identity element [13].

and remember that d represents the degree to which x includes y, which generalizes the algebraic inclusion $\leq$. [3] This can be compared to Cox's notation where instead of the function $\phi(x,y)$, he writes $(y \to x)$ for the special case where he is considering implication among logical statements. By replacing the comma with the solidus $\phi(x|y)$, we obtain a notation more reminiscent of probability theory.

Now consider two join-irreducible elements of the lattice, a and b, where $a \wedge b = \perp$, and a third element t such that $a \leq t$ and $b \leq t$. We will consider the degree to which the join $a \vee b$ of these join-irreducible elements includes the element t. As $a \vee b$ is itself a lattice element, the function ϕ allows us to describe the degree to which $a \vee b$ includes t. This degree is written as $\phi(a \vee b, t)$. As $a \wedge b = \perp$, this degree of inclusion can only be a function of $\phi(a,t)$ and $\phi(b,t)$, which can be written as

$$\phi(a \vee b, t) = S(\phi(a,t), \phi(b,t)). \tag{2}$$

Our goal is to determine the function S, which will tell us how to use the degree to which a includes t and the degree to which b includes t to compute the degree to which $a \vee b$ includes t. In this sense we are extending the algebra to a calculus.

The function S must maintain consistency with the lattice structure, or equivalently with its associated algebra. If we now consider another join-irreducible element $c \leq t$ where $a \wedge c = \perp$ and $b \wedge c = \perp$, and form the lattice element $(a \vee b) \vee c$, we can use associativity of the lattice to write this element a second way

$$(a \vee b) \vee c = a \vee (b \vee c). \tag{3}$$

Consistency with associativity requires that each expression gives exactly the same result

$$S(\phi(a \vee b, t), \phi(c,t)) = S(\phi(a,t), \phi(b \vee c, t)). \tag{4}$$

Applying S to the arguments $\phi(a \vee b, t)$ and $\phi(b \vee c, t)$ above, we get

$$S(S(\phi(a,t), \phi(b,t)), \phi(c,t)) = S(\phi(a,t), S(\phi(b,t), \phi(c,t))). \tag{5}$$

This can be further simplified by letting $u = \phi(a,t)$, $v = \phi(b,t)$, and $w = \phi(c,t)$, which gives

$$S(S(u,v), w) = S(u, S(v,w)). \tag{6}$$

This result is an equation for the function S, which emphasizes its property of associativity. To people who are familiar with Cox's work [3, 4], this functional equation should be immediately recognizable as what Aczél appropriately called *the associativity equation* [12, pp.253-273]. In Cox's derivation of probability theory we are accustomed to seeing this in terms of the logical conjunction. However, it is important to recognize that both the conjunction and disjunction follow associativity, and that this result generalized to the join is perfectly reasonable. The general solution, from Aczél [12], is

$$S(u,v) = f(f^{-1}(u) + f^{-1}(v)), \tag{7}$$

---

[3] This diverges from Caticha's development as he considers functions that take a single poset element as its argument. I discuss this difference in more detail in the sections that follow.

where f is an arbitrary function. This can be simplified by letting $g = f^{-1}$

$$g(S(u,v)) = g(u) + g(v),\tag{8}$$

and writing this in terms of the original lattice elements we find that

$$g(\phi(a \vee b,t)) = g(\phi(a,t)) + g(\phi(b,t)).\tag{9}$$

As Caticha emphasizes, this result is remarkable, as it reveals that there exists a function $g : \mathbb{R} \to \mathbb{R}$ re-mapping these numbers to a more convenient representation. Thus we can define a new map from the lattice elements to the real numbers, such that $v(a,t) \equiv g(\phi(a,t))$. This lets us write the combination rule as

$$v(a \vee b,t) = v(a,t) + v(b,t),\tag{10}$$

which is the familiar *sum rule* of probability theory for mutually exclusive (join-irreducible) assertions a and b

$$p(a \vee b|t) = p(a|t) + p(b|t).\tag{11}$$

This result is extremely important, as I have made no reference at all to probability theory in the derivation. Only consistency with associativity of the join, which is a property of all lattices, has been used. This means that for a given lattice, we can define a mapping from a pair of its elements to a real number, and when we take joins of its join-irreducible elements, we can compute the new value of this join by taking the sum of the two numbers. There are some interesting restrictions, which I will discuss later.

Extending the Sum Rule

The assumption made above was that the two lattice elements were join-irreducible and that their meet was the bottom element. How do we perform the computation in the event that this is not the case? In this section I will demonstrate how the sum rule can be extended in a distributive lattice. Consider two lattice elements x and y. In a distributive lattice all elements can be written as a unique join of join-irreducible elements

$$x = \left(\bigvee_{i=1}^{n} p_i\right) \vee \left(\bigvee_{i=1}^{k} q_i\right)\tag{12}$$

and

$$y = \left(\bigvee_{i=1}^{m} r_i\right) \vee \left(\bigvee_{i=1}^{k} q_i\right),\tag{13}$$

where I have written them so that the join-irreducible elements they have in common are the elements $q_1, q_2, \dots q_k$. The join of x and y can be written as

$$x \vee y = \left(\bigvee_{i=1}^{n} p_i\right) \vee \left(\bigvee_{i=1}^{k} q_i\right) \vee \left(\bigvee_{i=1}^{m} r_i\right) \vee \left(\bigvee_{i=1}^{k} q_i\right),\tag{14}$$

which can be simplified to

$$x \vee y = (\bigvee_{i=1}^{n} p_i) \vee (\bigvee_{i=1}^{k} q_i) \vee (\bigvee_{i=1}^{m} r_i), \tag{15}$$

since by $L1$ and $L2$

$$(\bigvee_{i=1}^{k} q_i) \vee (\bigvee_{i=1}^{k} q_i) = \bigvee_{i=1}^{k} q_i. \tag{16}$$

Since the p_i, q_i, and r_i are all join-irreducible elements, we can use the sum rule repeatedly to obtain

$$v(x \vee y, t) = \sum_{i=1}^{n} v(p_i, t) + \sum_{i=1}^{k} v(q_i, t) + \sum_{i=1}^{m} v(r_i, t). \tag{17}$$

Notice that the first two terms on the right are $v(x, t)$. I will add two more terms to the right (which cancel one another), and then group the terms conveniently

$$v(x \vee y, t) = (\sum_{i=1}^{n} v(p_i, t) + \sum_{i=1}^{k} v(q_i, t)) + (\sum_{i=1}^{m} v(r_i, t) + \sum_{i=1}^{k} v(q_i, t)) - \sum_{i=1}^{k} v(q_i, t). \tag{18}$$

This can be further simplified to give

$$v(x \vee y, t) = v(x, t) + v(y, t) - \sum_{i=1}^{k} v(q_i, t), \tag{19}$$

where we have the original sum rule minus a cross-term of sorts, which avoids double-counting the join-irreducible elements. The lattice elements forming this additional term can be found from

$$x \wedge y = \bigvee_{i=1}^{k} q_i, \tag{20}$$

so that

$$v(x \wedge y, t) = \sum_{i=1}^{k} v(q_i, t). \tag{21}$$

Note that to maintain consistency with the original sum rule, we must require that

$$v(\bot, t) = 0. \tag{22}$$

This allows us to write the generalized sum rule as

$$v(x \vee y, t) = v(x, t) + v(y, t) - v(x \wedge y, t). \tag{23}$$

What is remarkable is that we have derived a result that is valid for all distributive lattices! When joins of more than two elements are considered, this procedure can be iterated to avoid double-counting the join-irreducible elements that the elements share. Changing notation slightly this equation is identical to the general sum rule for probability theory.

$$p(x \vee y | t) = p(x | t) + p(y | t) - p(x \wedge y | t). \tag{24}$$

Valuations

I recently found that these ideas have been developing independently in geometry and combinatorics with influence of Gian-Carlo Rota [14, 13, 15]. A *valuation* is *defined* on a lattice of sets (distributive lattice) as a function $v : L \to \mathbb{A}$ that takes a lattice element to an element of a commutative ring with identity $\mathbb{A}$, and satisfies

$$v(a \vee b) + v(a \wedge b) = v(a) + v(b), \tag{25}$$

where $a, b \in L$, and

$$v(\bot) = 0. \tag{26}$$

By subtracting $v(a \wedge b)$ from both sides, we see that the valuation equation is just the generalized sum rule (Eqn. 23)

$$v(a \vee b) = v(a) + v(b) - v(a \wedge b). \tag{27}$$

When applied to Boolean lattices, valuations are called measures. As far as I am aware, the valuation equation was defined by mathematicians and not derived. However, following Cox and Caticha, we have derived it directly from the sum rule, which was derived from consistency with associativity of the join.

Note that valuations, as we describe them here have a single argument and do not explicitly consider the degree to which one lattice element includes a second lattice element. This is not a problem, as one can define bi-valuations, tri-valuations, and multi-valuations although it is not clear how to interpret all of these functions as generalizations of inclusion on a poset. One can define valuations that represent the degree to which an element x includes $\top$, by

$$v(x) \equiv v(x, \top), \tag{28}$$

which can be interpreted in probability theory as a prior probability [4]

$$v(x) \equiv p(x|\top). \tag{29}$$

However, throughout this paper I will work with bi-valuations, as they can be used to represent the degree to which one lattice element includes another.

Möbius Functions

I have demonstrated how the sum rule can be extended for distributive lattices, but how is this handled in general for posets where associativity holds? One must rely on

---

[4] In this notation $\top$ refers to the truism, which is the join of all possible statements. In the past, I have preferred to write probabilities in the style of Jaynes where I is used to represent our 'prior information' as in $p(x|I)$. The truism represents this prior information in part, since we know *a priori* that one of the exhaustive set of mutually exclusive assertions is true. However, excluded in the notation $p(x|\top)$ is explicit reference to the part of the prior information I that is relevant to the probability assignments. I choose to leave I out of the probability notation here to emphasize the fact that the function p takes two lattice elements as arguments—not abstract symbols like I.

what is called the Möbius function for the poset. I will begin by discussing a special class of real-valued functions of two variables defined on a poset, such as $f(x,y)$, which are non-zero only when $x \leq y$. This set of functions comprises the *incidence algebra* of the poset [16]. The sum of two functions $h = f + g$ is defined the usual way by

$$h(x,y) = f(x,y) + g(x,y), \tag{30}$$

as is multiplication by a scalar $h = \lambda f$. However, the product of two functions in the incidence algebra is found by taking the convolution over the interval of elements in the poset

$$h(x,y) = \sum_{x \leq z \leq y} f(x,z)g(z,y). \tag{31}$$

We can define three useful functions [16, 17]

$$\delta(x,y) \;=\; \begin{cases} 1 & if\ x = y \\ 0 & if\ x \neq y \end{cases} \quad (\textit{Kronecker delta function}) \tag{32}$$

$$n(x,y) \;=\; \begin{cases} 1 & if\ x < y \\ 0 & if\ x \nless y \end{cases} \quad (\textit{incidence function}) \tag{33}$$

$$\zeta(x,y) \;=\; \begin{cases} 1 & if\ x \leq y \\ 0 & if\ x \nleq y \end{cases} \quad (\textit{zeta function}) \tag{34}$$

The delta function indicates when two poset elements are equal. The incidence function indicates when an element x is properly contained in an element y. Last, the zeta function indicates whether y includes x, which means that the zeta function is equal to the sum of the delta function and the incidence function

$$\zeta(x,y) = n(x,y) + \delta(x,y). \tag{35}$$

It is important to be able to invert functions in the incidence algebra. For example, the inverse of the zeta function, $\mu(x,y)$ satisfies

$$\sum_{x \leq z \leq y} \zeta(x,z)\mu(z,y) = \delta(x,y). \tag{36}$$

One can show [13, 16, 18] that the function $\mu(x,y)$, called the *Möbius function*, is defined by

$$\begin{aligned} \mu(x,x) &= 1 & x \in P \\ \sum_{x \leq z \leq y} \mu(x,z) &= 0 & x < y \\ \mu(x,y) &= 0 & if\ x \nleq y, \end{aligned} \tag{37}$$

where

$$\sum_{x \leq z \leq y} \mu(x,z) = \sum_{x \leq z \leq y} \mu(z,y). \tag{38}$$

Rather than providing a proof, I will demonstrate this by considering the possible cases. Obviously, if $x = y$ then

$$\zeta(y,y)\mu(y,y) = 1, \tag{39}$$

which is consistent with the first condition for the Möbius function (37). Next, if $x \leq y$ we can use the fact that $\zeta(x,z) = 1$ only when $x \leq z$ to rewrite (36)

$$\sum_{x \leq z \leq y} \zeta(x,z)\mu(z,y) = \delta(x,y). \tag{40}$$

as

$$\sum_{x \leq z \leq y} \mu(z,y) = \delta(x,y). \tag{41}$$

The sum can be rearranged using (38)

$$\sum_{x \leq z \leq y} \mu(x,z) = \delta(x,y), \tag{42}$$

which is consistent with the second condition (37) when $x < y$. Last, if $x > y$ then (36) is trivially satisfied.

More importantly, the Möbius function is used to invert valuations on a poset P [13, 16, 18] so that given a function

$$g(x) = \sum_{d \leq x} f(d) \tag{43}$$

one can find $f(x)$ by

$$f(x) = \sum_{d \leq x} \mu(d,x)g(d). \tag{44}$$

What's going on here is made more clear [18] by considering the poset of natural numbers N in the usual order (see Fig 1a). This poset is totally ordered and the Möbius function is easily found to be

$$\mu_N(x,y) = \begin{cases} 1 & if \ x = y \\ -1 & if \ x \prec y \\ 0 & otherwise \end{cases} \tag{45}$$

Given

$$g(n) = \sum_{m \leq n} f(m) \tag{46}$$

we find, using (44) and (45), that

$$f(n) = g(n) - g(n-1). \tag{47}$$

This is the finite difference operator, which is the discrete analog of the *fundamental theorem of calculus* for a poset [16, 18], which basically relates sums to differences.

Readers well-versed in number theory have seen the classic Möbius function [19] used to invert the Riemann zeta function

$$\zeta(s) = \sum_{n=1}^{\infty} \frac{1}{n^s}, \tag{48}$$

which is important in the study of prime numbers (refer to the poset in Fig. 1b). The Möbius function in this case is defined [16, 18, 7] as

$$\mu(1) = 1$$
$$\sum_{d|m} \mu(d) = 0 \tag{49}$$

where the sum is over all numbers d dividing m, that is all numbers $d \le m$ in the poset in Fig. 1b. Specifically, it values are found by

$$\mu(d) = \begin{cases} 0 & if \ d \ is \ divisible \ by \ some \ p^2 \\ (-1)^k & if \ is \ a \ product \ of \ k \ distinct \ primes \end{cases} \tag{50}$$

where p is a prime. This leads to the inverse of the zeta function given by

$$\frac{1}{\zeta(s)} = \sum_{n=1}^{\infty} \frac{\mu(n)}{n^s}. \tag{51}$$

Clearly, order theory and the incidence algebra ties together areas of mathematics such as the calculus of finite differences and number theory. In the next section, I will show that we can use the Möbius function to extend the sum rule over the entire poset.

The Inclusion-Exclusion Principle

By iterating the generalized sum rule, we obtain what Rota calls the *inclusion-exclusion principle* [15, p.7]

$$v(x_1 \vee x_2 \vee \cdots \vee x_n, t) = \sum_i v(x_i, t) - \sum_{i<j} v(x_i \wedge x_j, t) + \sum_{i<j<k} v(x_i \wedge x_j \wedge x_k, t) - \cdots \tag{52}$$

This equation holds for distributive lattices where every lattice element can be written as a unique join of elements. As I will show, this principle appears over and over again, and is a sign that order-theoretic principles and distributivity underlie the laws having this form.

Rota showed that the inclusion-exclusion principle can be obtained from the Möbius function of the poset [16, 18]. For a Boolean lattice of sets the Möbius function is given by

$$\mu_B(x, y) = (-1)^{|y| - |x|} \tag{53}$$

whenever $x \subseteq y$ and 0 otherwise, where $|x|$ is the cardinality of the set x. The Möbius function for a distributive lattice is similar

$$\mu_D(x, y) = \begin{cases} 1 & if \ x = y \\ (-1)^n & if \ x < y \\ 0 & if \ x \nleq y \end{cases} \tag{54}$$

where in the second case y is the join of n distinct elements covering x. This leads directly to the alternating sum and difference in the inclusion-exclusion principle as one sums down the lattice.

The inclusion-exclusion principle appears in a delightful variety of contexts, several of which will be explored later. We have already seen it in the context of the sum rule of probability theory

$$p(x \vee y|t) = p(x|t) + p(y|t) - p(x \wedge y|t). \tag{55}$$

It also appears in Cox's generalized entropies [4], which were explored earlier in McGill's multivariate generalization of mutual information [20], and examined more recently as co-information by Tony Bell [21]. The familiar mutual information makes the basic point

$$I(x,y) = H(x) + H(y) - H(x,y). \tag{56}$$

Again, this is the inclusion-exclusion principle at work. Rota [13] gives an interesting example from Pólya and Szegö [22, Vol II., p. 121] which I will shorten here to

$$max(a,b) = a + b - min(a,b), \tag{57}$$

where a and b are real numbers. This equation may seem horribly obvious, but like the others, it can be extended by iterating. The idea is to include and exclude all the way down the lattice! The inclusion-exclusion pattern is an important clue that order theory plays an important role.

Assigning Valuations

There are some interesting and useful results on assigning valuations and extending them to larger lattices (eg. [15, p.9]). The fact that probabilities are valuations implies that these results are relevant to assigning probabilities. Specifically, Rota [13, Theorem 1, Corollary 2, p.35] showed that

> A valuation in a finite distributive lattice is uniquely determined by the values it takes on the set of join-irreducibles of L, and these values can be arbitrarily assigned.

By considering $p(x|\top)$, this result can be applied to the assignment of prior probabilities. In the lattice of logical assertions ordered by logical implication, the join-irreducible elements are the exhaustive set of mutually exclusive assertions. Thus by assigning their prior probabilities, the probabilities of their various disjunctions are uniquely determined. This is easy, just use the sum rule.

More profound is the fact that Rota's theorem states that these assignments can be *arbitrary*. This means that there is no information in the Boolean algebra of these assertions, and hence the inferential calculus, to guide us in these assignments. Thus probability theory tells us nothing about assigning priors. Other principles, such as symmetry, constraints, and consistency with other aspects of the problem *must* be employed to assign priors. Once the priors are assigned, order-theoretic principles dictate the remaining probabilities through the inferential calculus.

Last, it is not clear to me how to assign valuations in posets where at least one element can be written as the join of join-irreducible elements in more than one way, such as in the lattice N_5 [9, Fig. 4.3(i)]. Once again, consistency must be the guiding principle. When there is more than one way to write an element as a join of join-irreducible elements, the valuations assigned to those elements must be consistent with the particular sum rule for that lattice. This is not an issue for probability theory, but it will become an issue when this methodology is extended to other problems.

Meets and the Product Rule

The product rule is usually seen as being necessary for computing probabilities of conjunctions of logical statements, whereas the sum rule is necessary for computing the probabilities of disjunctions. This actually isn't true. The sum rule allows one to compute the probabilities of conjunctions equally well. Rearranging Equation 24 gives

$$p(x \wedge y|t) = p(x|t) + p(y|t) - p(x \vee y|t), \tag{58}$$

or more generally

$$v(x \wedge y, t) = v(x, t) + v(y, t) - v(x \vee y, t). \tag{59}$$

The important point is that for some problems, this just isn't useful.

The key to understanding the product rule is to realize that there are actually two kinds of logical conjunctions in probability theory. The first is the 'and' that is implemented by the meet on the Boolean lattice. In the earlier example on who stole the tarts, this type of conjunction leads to statements like $(k \vee a) \wedge (n \vee a)$, which can be read literally as '*Somebody stole the tarts **and** it wasn't the Knave!*' The meet performs the role of the logical conjunction while working *within* the hypothesis space. The second type of logical conjunction occurs when one *combines* two hypothesis spaces to make a bigger space. For example, we can combine a lattice F describing different types of fruit, with a lattice Q describing the quality of the fruit by taking a Cartesian product of $F \times Q$, which results in statements like '*The fruit is an apple **and** it is spoiled!*' This type of conjunction is not readily computed using the sum rule. To handle both types of logical conjunctions, we will derive the product rule. In the section on quantum mechanics, we will see that these ideas are not limited to probability theory.

The Lattice Product

We can define the lattice F describing the type of fruit by specifying the two atomic assertions covering the bottom

$$a = \text{'It is an apple!'}$$
$$b = \text{'It is a banana!'}.$$

This will form a Boolean lattice with four elements shown on the left-side of Figure 2a. This Boolean lattice structure with two atoms is denoted by 2^2. As usual, the top element

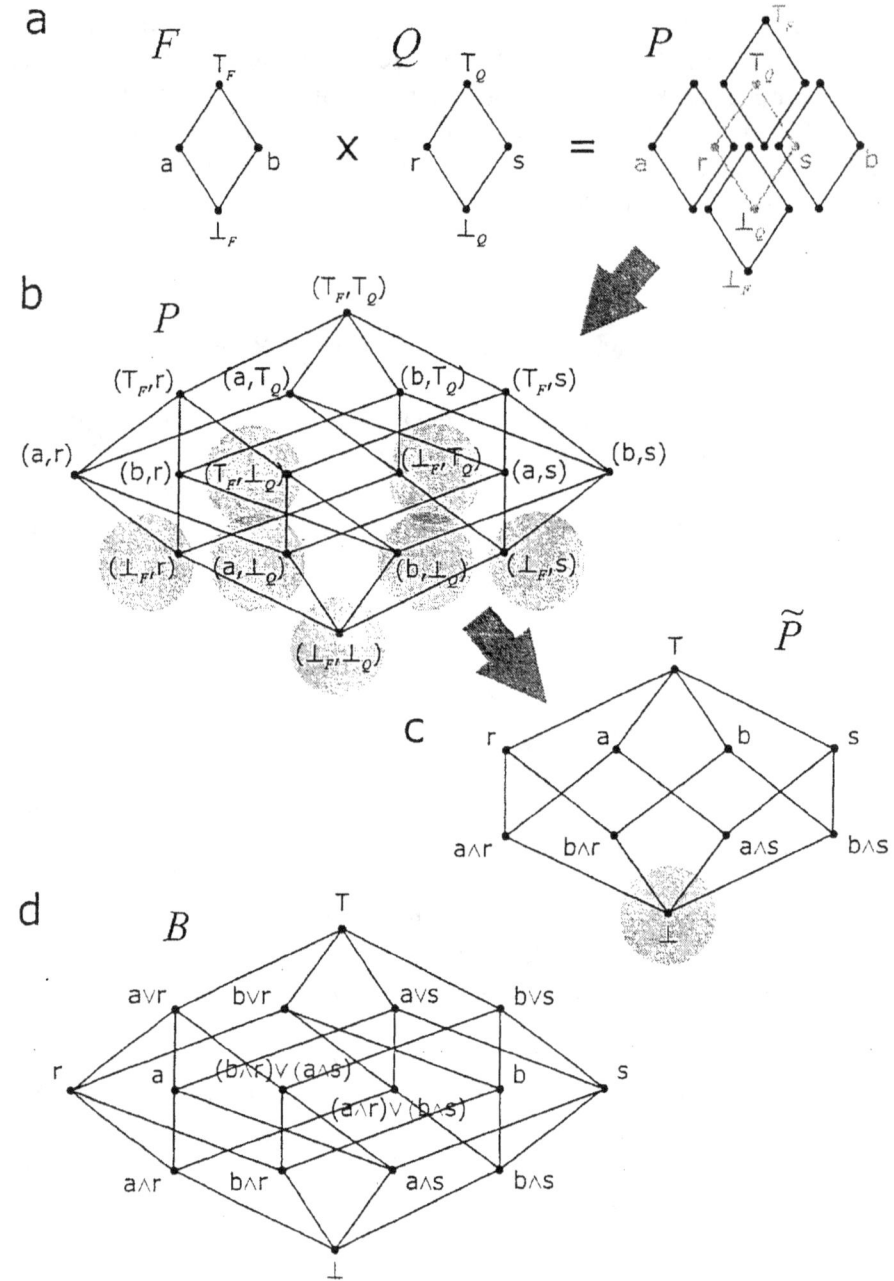

FIGURE 2. a. The product of two Boolean lattices F and Q can be constructed by taking the Cartesian product of their respective elements. b. This leads to the product lattice $P = F \times Q$, which represents both spaces jointly. However, P contains seven statements (gray circles), which belong to an equivalence class of absurd statements. c. By grouping these elements in the equivalence class under $\perp$, we can form a new lattice $\tilde{P}$, which is a subdirect product of F and Q. d. An alternate lattice structure can be formed by treating the statements represented by the join-irreducible elements of the subdirect product $\tilde{P}$ as atomic statements and forming their powerset. This yields a Boolean lattice B distinct from, yet isomorphic to the Cartesian product P. Both logical structures $\tilde{P}$ and B are implicitly used in probability theory, and both follow the distributive law $D1$.

stands for the truism, which is a logical statement that says *'It is an apple or a banana!'*, which we write symbolically as $\top_F = a \vee b$. The bottom is the absurdity which says *'It is an apple and a banana!'*, written $\bot_F = a \wedge b$. The lattice is clearly Boolean as the complement of a is b, and vice versa (i.e. $a \wedge b = \bot_F$ and $a \vee b = \top_F$).

Similarly, what is known about the quality of the fruit can be described by the lattice Q generated by the atomic assertions

$$r = \text{'It is ripe!'}$$
$$s = \text{'It is spoiled!'}.$$

These assertions generate the center lattice in Figure 2a, which is also Boolean (isomorphic to 2^2).

We can combine these two lattices by taking the *lattice product*. Graphically, this can be constructed by fixing one of the two lattices, and placing copies of the other over each element of the former. In Figure 2a, I fix the lattice Q and place copies of the lattice F over each element of Q. The elements of the product lattice $P = F \times Q$ are found by forming the Cartesian product of the elements of the two lattices. For example, the element (a, r) represents the logical statement that says *'The fruit is an apple and it is ripe!'* Two elements of $F \times Q$, (f_1, q_1) and (f_2, q_2), can be ordered coordinate-wise [9, p.42], so that

$$(f_1, q_1) \leq (f_2, q_2) \quad \text{when} \quad f_1 \leq f_2 \quad \text{and} \quad q_1 \leq q_2, \tag{60}$$

which leads to a coordinate-wise definition of $\vee$ and $\wedge$

$$(f_1, q_1) \vee (f_2, q_2) = (f_1 \vee f_2, q_1 \vee q_2) \tag{61}$$

$$(f_1, q_1) \wedge (f_2, q_2) = (f_1 \wedge f_2, q_1 \wedge q_2). \tag{62}$$

The result is the Boolean lattice P in Figure 2b, which is isomorphic to $2^2 \times 2^2 = 2^4$. I should make the meaning of some of these statements more explicit. For example, $(\top_F, r)$ is a statement that says *'It is an apple or a banana and it is ripe!'* Similarly, $(b, \top_Q)$ is a statement that says *'It is a banana that is either ripe or spoiled!'*

The lattice product is associative, so that for three lattices J, K, and L

$$J \times (K \times L) = (J \times K) \times L. \tag{63}$$

This associativity translates to the associativity of the meet of the Cartesian products of the elements, which is consistent with the fact that the lattice product of two lattices is a lattice.

What is interesting is that there are seven elements (gray circles) that involve at least one of the two absurdities. These seven elements, $(\top_F, \bot_Q)$, $(\bot_F, \top_Q)$, $(\bot_F, r)$, $(a, \bot_Q)$, $(b, \bot_Q)$, $(\bot_F, s)$, and $(\bot_F, \bot_F)$ each belong to an *equivalence class* of logically absurd statements as they say things like *'It is ripe and spoiled!'* and *'It is an apple and a banana!'* If we group these absurd statements together under a new symbol $\bot$, we can construct the lattice $\tilde{P}$ in Figure 2c. I have simplified the element labels so that the element b stands for $(b, \top_Q)$, which can be read as *'The fruit is a banana!'*, and s stands for $(\top_F, s)$, which can be read as *'The fruit is spoiled!'* The lattice $\tilde{P}$ is a *subdirect product* of the lattices F and Q since $\tilde{P}$ can be embedded into their Cartesian product P, and its projections onto both F and Q are surjective (onto, but not one-to-one).

The symbol $\wedge$ again represents the meet in the subdirect product $\tilde{P}$, so that the meet of a and s gives the element $a \wedge s$, which is equivalent to a spoiled apple (a, s). In this way we see that the meet in the equivalent product lattice behaves like the meet in the original hypothesis space, while simultaneously implementing the Cartesian product. However, there are some key differences. First, $\tilde{P}$ is not Boolean. For example, there is no unique complement to the statement $a \wedge s$, as both b and r satisfy the requirements for the complement

$$(a \wedge s) \vee b = \top$$
$$(a \wedge s) \wedge b = \bot$$

and

$$(a \wedge s) \vee r = \top$$
$$(a \wedge s) \wedge r = \bot$$

A more important difference is the fact that the meet in P follows both distributive laws $D1$ and $D2$, whereas the meet in $\tilde{P}$ follows only $D1$. To demonstrate this consider the following in the context of $\tilde{P}$

$$a \wedge (r \vee s) = a \wedge \top = a, \tag{64}$$

and

$$(a \wedge r) \vee (a \wedge s) = a, \tag{65}$$

which is consistent with $D1$

$$a \wedge (r \vee s) = (a \wedge r) \vee (a \wedge s). \tag{66}$$

Now consider

$$a \vee (r \wedge s) = a \vee \bot = a, \tag{67}$$

whereas

$$(a \vee r) \wedge (a \vee s) = \top \wedge \top = \top, \tag{68}$$

which is inconsistent with $D2$, which is distributivity of $\vee$ over $\wedge$. The difficulty here is clear. In the Cartesian product P and the subdirect product $\tilde{P}$, statements like $a \vee r$ and $a \vee s$ do not have distinct interpretations since one cannot use the Cartesian product to form the statement '*It is an apple or the fruit is ripe!*' Another way to look at the loss of property $D2$ is to notice that because we have identified the bottom elements in the equivalence relation we lose distributivity of $\vee$ over $\wedge$ (property $D2$), and maintain distributivity of $\wedge$ over $\vee$ (property $D1$). Had we identified the top elements we would have obtained the dual situation.

There is one last important construction we can perform. We can construct another lattice by considering the join-irreducible statements of the subdirect product differently. If we let $a \wedge r$, $a \wedge s$, $b \wedge r$, $b \wedge s$ represent an atomic set of exhaustive mutually exclusive statements rather than a Cartesian product of statements, then all other elements in the lattice can be formed from the joins of these atoms. The result is a Boolean lattice B that is isomorphic to the lattice product P, so that $B \sim P \sim \mathbf{2}^4$. In this lattice the atomic statements, such as $a \wedge r$, do not represent Cartesian products, but instead represent elementary statements like '*It is a ripe apple!*'. For this reason, B contains logical statements not included in the Cartesian product P or the subdirect product $\tilde{P}$. For example, we can construct statements like $(b \wedge r) \vee (a \wedge s)$, which state '*It is either a ripe banana or a spoiled apple!*' Lattices formed this way naturally follow both $D1$ and $D2$, as they are Boolean.

The important point here is that we use each of these constructs in different applications of probability theory without explicit consideration. Most relevant is the fact that each of these three lattice constructs P, $\tilde{P}$, and B follows $D1$. Thus if we require that our calculus satisfies distributivity of $\wedge$ over $\vee$ then we will have a rule that is consistent with each of the logical constructs we have considered here.

Deriving the Product Rule

The product rule is important as it gives us a way to compute the degree of inclusion for meets of elements in a lattice constructed from the product of two distributive lattices. We look for a function P that allows us to write the degree to which the meet of two elements $x \wedge y$ includes a third element t without relying on the join $x \vee y$. Cox chose the form

$$v(x \wedge y, t) = P(v(x,t), v(y, x \wedge t)), \tag{69}$$

which was later shown by Tribus [23] and Smith & Erickson [24] to be the only functional form that satisfies consistency with associativity of $\wedge$. [5]

Consider the elements a and b where $a \wedge b = \perp$, and the elements r and s where $r \wedge s = \perp$. We reproduce Caticha's derivation [11] and consider distributivity $D1$ of the meet over the join in the lattice product

$$(a, (r \vee s)) \equiv a \wedge (r \vee s) = (a \wedge r) \vee (a \wedge s). \tag{70}$$

This equation gives us two different ways to express the same poset element. Consistency with distributivity $D1$ requires that the same value is associated with each of these two expressions. Using the sum rule (24) and the form of P consistent with associativity (69), we find that distributivity requires that

$$P(v(a,t), v(r \vee s, a \wedge t)) = v(a \wedge r, t) + v(a \wedge s, t), \tag{71}$$

which further simplifies to

$$P(v(a,t), v(r, a \wedge t) + v(s, a \wedge t)) = P(v(a,t), v(r, a \wedge t)) + P(v(a,t), v(s, a \wedge t)). \tag{72}$$

If we let $u = v(a,t)$, $v = v(r, a \wedge t)$, and $w = v(s, a \wedge t)$, the equation above can be written more concisely as

$$P(u, v + w) = P(u, v) + P(u, w). \tag{73}$$

---

[5] I use the sum rule in the following derivation, which requires that I use the mapping v rather than ϕ.

This is a functional equation for the function P, which captures the essence of distributivity. By working with this equation, we will obtain the functional form of P.

The idea is to show that $P(u, v+w)$ is linear in its second argument. If we let $z = w+v$, and write (73) as

$$P(u,z) = P(u,v) + P(u,w), \tag{74}$$

we can look at the second derivative with respect to z. Using the chain rule we find that

$$\frac{\partial}{\partial v} = \frac{\partial z}{\partial v}\frac{\partial}{\partial z} = \frac{\partial}{\partial z}. \tag{75}$$

This can be done also for w giving

$$\frac{\partial}{\partial v} = \frac{\partial}{\partial w} = \frac{\partial}{\partial z}. \tag{76}$$

Writing the second derivative with respect to z as

$$\frac{\partial^2}{\partial z^2} = \frac{\partial}{\partial v}\frac{\partial}{\partial w}, \tag{77}$$

we find that

$$
\begin{aligned}
\frac{\partial^2}{\partial z^2}P(u,z) &= \frac{\partial}{\partial v}\frac{\partial}{\partial w}(P(u,v) + P(u,w)) \\
&= \frac{\partial}{\partial y}\left(\frac{\partial}{\partial w}P(u,w)\right) \\
&= \frac{\partial}{\partial w}\left(\frac{\partial}{\partial v}P(u,w)\right) \\
&= 0,
\end{aligned}
\tag{78}
$$

which means that the function P is linear in its second argument.

$$P(u,v) = A(u)v + B(u). \tag{79}$$

If (79) is substituted back into (73) we find that $B(u) = 0$.

We can use $D1$ another way by considering $(a \vee b) \wedge r$. This leads to a condition that looks like

$$P(v+w, u) = P(v, u) + P(w, u), \tag{80}$$

where u, v, w are appropriately redefined. Following the approach above, we find that P is also linear in its first argument

$$P(u,v) = A(v)u. \tag{81}$$

Together with (79), the general solution is

$$P(u,v) = Cuv, \tag{82}$$

where C is an arbitrary constant. Thus, we have the *product rule*

$$v(x \wedge y, t) = Cv(x,t)v(y, x \wedge t), \tag{83}$$

which tells us the degree to which the new element $x \wedge y$ includes the element t. This looks more familiar if we set $C = 1$ and re-write the rule in probability-theoretic notation,

$$p(x \wedge y | t) = p(x|t)p(y|x \wedge t). \tag{84}$$

If the lattice we are working in is formed by the lattice product, (83) can be rewritten using the Cartesian product notation

$$v((x,y),(t_x,t_y)) = Cv((x,\top_y),(t_x,t_y))v((\top_x,y),(x \wedge t_x,t_y)), \tag{85}$$

where $x \wedge y \equiv (x,y)$, $t \equiv (t_x,t_y)$, $x \equiv (x,\top_y)$, and $y \equiv (\top_x,y)$. Simplifying, we see that

$$v((x,y),(t_x,t_y)) = Cv(x,t_x)v(y,t_y), \tag{86}$$

where $v(x,t_x)$ is computed in one lattice and $v(y,t_y)$ in the other.

BAYES' THEOREM

The origin of Bayes' Theorem is perhaps the most well-known. It derives directly from the product rule and consistency with commutativity of the meet. When it is true that $x \wedge y = y \wedge x$, one can write the product rule two ways

$$v(x \wedge y, t) = Cv(x,t)v(y,x \wedge t)$$

and

$$v(y \wedge x, t) = Cv(y,t)v(x,y \wedge t).$$

Consistency with commutativity requires that these two results are equal. Setting them equal and rearranging the terms leads to *Bayes' Theorem*

$$v(x, y \wedge t) = \frac{v(x,t)v(y,x \wedge t)}{v(y,t)}. \tag{87}$$

This will look more familiar if we let $x \equiv h$ be a logical statement representing a hypothesis, and let $y \equiv d$ be a new piece of data, and change notation to that used in probability theory

$$p(h | d \wedge t) = \frac{p(h|t)p(d|h \wedge t)}{p(d|t)}. \tag{88}$$

This makes more sense now that it is clear that the two statements h and d come from different lattices. This is why d represents a *new* piece of data, it represents information we were not privy to when we implicitly constructed the lattice including the hypothesis h. The statements h and d belong to two different logical structures and Bayes' Theorem tells us how to do the computation when we combine them. To perform this computation, we need to first make assignments. The prior assignment $p(h|t)$ is made in the original lattice of hypotheses, whereas the likelihood assignment $p(d|h \wedge t)$ is made in the product lattice. The evidence, while usually not assigned, refers to an assignment that would take place in the original data lattice.

LAWS FROM ORDER

Why is all this important? Because, the sum rules are associated with all lattices, and sum and product rules are not just associated with Boolean algebra, but with distributive algebras. This is a much wider range of application than was ever considered by Cox, as the following examples demonstrate.

Information Theory from Order .

Most relevant to Cox is the further development of the calculus of inquiry [25, 26, 27, 8, 6], which appears to be a natural generalization of information theory. The extension of Cox's methodology to distributive lattices in general is extremely important to this development, as the lattice structure of questions is not a Boolean lattice, but is instead the *free distributive lattice* [8, 6]. This free distributive lattice of questions is generated from the ordered set of down-sets of its corresponding Boolean lattice of assertions. A probability measure on the Boolean assertion lattice induces a valuation, which we call *relevance* or *bearing*, on the question lattice. I will show in a future paper [28] that the join-irreducible questions, called *elementary questions* by Fry [26], have relevances that are equal to $-p \log p$, called the *surprise*, where p is the probability of the assertion defining the elementary question. Joins of these elementary questions form real questions, which via the sum rule have valuations formed from sums of surprises, or *entropy*. Going further up the question lattice, one uses the generalized sum rule, which results in mutual information, and eventually generalized entropy [4], also called co-information [21]. The lattice structure is given by the algebra of questions, and the generalized entropies are the valuations induced on the question lattice by the probability measure assigned to the Boolean algebra of assertions. Exactly how these valuations are induced, manipulated, and used in calculations regarding questions will be discussed elsewhere [28].

Geometry from Order

Perhaps more interesting is the fact that much of geometry can be derived from these order-theoretic concepts. There has been much work done in this area, which is called *geometric probability*. This invariably calls up thoughts of Buffon's Needle problem, but the range of applications is much greater. I have found the introductory book by Klain and Rota [15] to be very useful, and below I discuss several illustrative examples from their text.

The Lattice of Parallelotopes

Imagine a Cartesian coordinate system in n-dimensions. We will consider orthogonal parallelotopes, which are rectangular boxes with sides parallel to the coordinate axes.

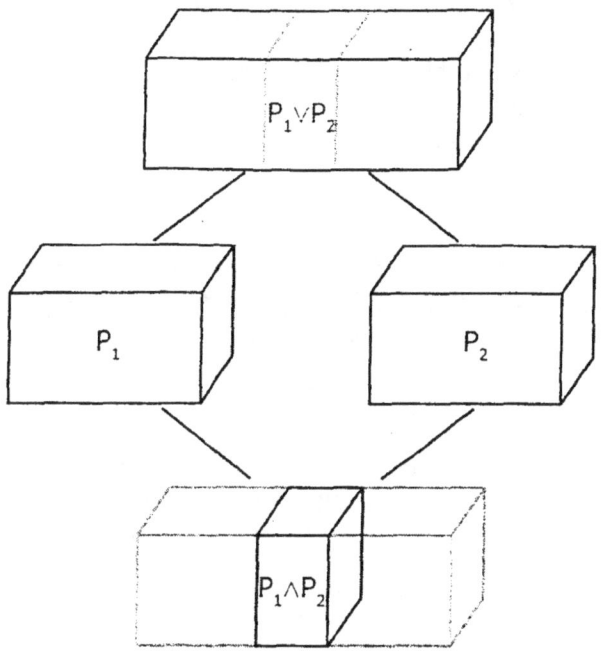

FIGURE 3.   An illustration of the join and meet of two orthogonal parallelotopes.

By taking finite unions (joins) and intersections (meets) of these parallelotopes, we can construct the lattice (or equivalently the algebra) of n-dimensional parallelotopes $Par(n)$ [15, p.30]. In one dimension, the parallelotope is simply a closed interval on the real line.

To extend the algebra to a calculus, to each parallelotope P we will assign a valuation[6] $\mu(P)$, which is invariant with respect to geometric symmetries relevant to $Par(n)$, such as invariance with respect to translations along the coordinate system and permutations of the coordinate labels. Assigning valuations that are consistent with these important geometric symmetries is analogous to Jaynes' use of group invariance to derive prior probabilities [29].

Before looking at these invariant valuations in more detail, Figure 3 shows the join of two parallelotopes P_1 and P_2. Since this lattice is distributive, the sum rule can be used to compute the valuation of the join $P_1 \vee P_2$

$$\mu(P_1 \vee P_2) = \mu(P_1) + \mu(P_2) - \mu(P_1 \wedge P_2). \tag{89}$$

Again we see the familiar inclusion-exclusion principle.

---

[6] Note that one can consider $\mu(P) \equiv \mu(P, T)$, where T refers to a parallelotope to which the measure is in some sense normalized.

A Basis Set of Invariant Valuations

At this point, you have probably already identified a valuation that will satisfy invariance with respect to translation and coordinate label permutation. One obvious valuation for three-dimensional parallelotopes suggested by the illustration is *volume*

$$\mu_3(x) = x_1 x_2 x_3, \tag{90}$$

where x_1, x_2 and x_3 are the side-lengths of the parallelotope. Surprisingly, this is not the only valuation that satisfies the invariance properties we are considering. There is also a valuation, which is proportional to the *surface area*

$$\mu_2(x) = x_1 x_2 + x_2 x_3 + x_3 x_1, \tag{91}$$

which is easily shown to satisfy both the invariance properties as well as the sum rule. In fact, there is a basis set of invariant valuations, which in the case of three-dimensional parallelotopes consists of the volume, surface area, *mean width*

$$\mu_1(x) = x_1 + x_2 + x_3, \tag{92}$$

and the *Euler characteristic* μ_0, which for parallelotopes is equal to one for non-empty parallelotopes and zero otherwise. The fact that these valuations form a basis means that we can write any valuation as a linear combination of these basis valuations

$$\mu = a\mu_3 + b\mu_2 + c\mu_1 + d\mu_0. \tag{93}$$

In general, it is not clear under which invariance conditions one obtains a basis set of valuations rather than a unique functional form. This is an extremely important issue when we consider the issue of assigning prior probabilities in probability theory. Jaynes' demonstrated how to derive priors that are consistent with certain invariances, and cautioned that if the number of parameters in the transformation group is less than the number of model parameters, then the prior will only be partially determined [29]. How to consistently assign priors in this case is an open problem.

Furthermore, the Euler characteristic is interesting in that it takes on discrete rather than continuous values. This is something that is not seen in measure theory indicating that this development is more general than the typical measure-theoretic approaches. An important example of this has been identified within the context of information theory. Aczél [30] showed that the Hartley entropy [31], which takes on only discrete values, has certain 'natural' properties shared only with the Shannon entropy [32]. This will also be discussed in more detail elsewhere [28].

The Euler Characteristic

The Euler characteristic appears in other lattices and is perhaps best known from the lattice of convex polyhedra where it satisfies the following formula

$$\mu_0 = F - E + V, \tag{94}$$

where F is the number of faces, E is the number of edges, and V is the number of vertices [14]. For all convex polyhedra in three-dimensions, $\mu_0 = 2$. For example, if we consider a cube, we see that it has 6 faces, 12 edges, and 8 vertices, so that $6 - 12 + 8 = 2$. Again, this is an example of the inclusion-exclusion principle, which comes from the sum rule. In the lattice of simplicial complexes, a face is a 2-simplex, an edge is a 1-simplex, and a vertex is a 0-simplex. To compute the characteristic, we add at one level, subtract at the next, and add at the next and so on. This geometric law derives directly from order theory via the sum rule.

Spherical Triangles

The connections with order theory do not stop with polyhedra, but extend into continuous geometry. Klain & Rota [15, p.158] show that the solid angle subtended by a triangle inscribed on a sphere, called the *spherical excess*, can be found using the inclusion-exclusion principle

$$\Omega(\Delta) = \alpha + \beta + \gamma - \pi \tag{95}$$

where $\alpha, \beta, \gamma \in [0, \pi]$ denote the angles of the spherical triangle. Such examples highlight the degree to which order theory dictates laws.

Statistical Physics from Order

Up to this point, probability theory has been the main example by which I have demonstrated the use of order-theoretic principles to derive a calculus from an algebra. Thanks to the efforts of Ed Jaynes [33], Myron Tribus [34] and others, I am able to wave my hands and state that statistical physics derives from order-theoretic principles. In one important respect this argument is a sham, and that is where entropy is concerned. The *principle of maximum entropy* [33, 35], which is central to statistical physics, lies just beyond the scope of this order-theoretic framework. It is possible that a fully-developed calculus of inquiry [28] will provide useful insights. With entropies being associated with the question lattice, application of the principle of maximum entropy to enforce consistency with known constraints may in some sense be dual to the *maximum a posteriori* procedure in probability theory. However, at this stage it is certain that the probability-based features of the theory of statistical physics derive directly from these order-theoretic principles.

Quantum Mechanics from Order

Most surprising is Ariel Caticha's realization that the rules for manipulating quantum mechanical amplitudes in slit experiments derive from consistency with associativity and distributivity of experimental setups [11]. Each experimental setup describes an idealized experiment that describes a particle prepared at an initial point and detected at a final point. These experimental setups are simplified so that the only observable considered is that of position. The design of each setup constrains the statements that one can make about the particle.

These setups can be combined in two ways, which I will show are essentially *meets* and *joins* of elements of a poset. However, there are additional constraints on these operations imposed by the physics of the problem. We will see that the meets are not commutative, and this makes these algebraic operations significantly different from the AND and OR of propositional logic. This lack of commutativity means that there is no Bayes' Theorem analog for quantum mechanical amplitudes. Furthermore, the operation of negation is never introduced—nor is it necessary. This sets Caticha's approach apart from other quantum logic approaches where the negation of a quantum mechanical proposition remains a necessary concept.

Caticha considers a simple case where the only experiments that can be performed are those that detect the local presence or absence of a particle. He considers a particle, which is prepared at time t_i at position x_i and is later detected at time t_f at position x_f. Experimental setups of this sort will test statements like '*the particle moved from x_i to x_f*'. The physics becomes interesting when we place obstacles in the path of the particle. For example, we can place a barrier with a single slit at position x_1. The detector at x_f will only detect the particle if it moves from x_i to x_1 and then onward to x_f, where $t_i < t_1 < t_f$. [7] This barrier with a slit imposes a constraint on the particles that can be detected. The central idea is that experimental setups can be combined in two ways: increasing the number of constraints on the particle behavior, or by decreasing the number of constraints. This allows one to impose an ordering relation on the experimental setups, and by considering the set of all possible setups where the particle is prepared at x_i and detected at x_f we have a poset of experimental setups.

Setups with fewer constraints give greater freedom to the possible behavior of the detected particle. Of course, the ordering relation we choose can be defined in terms of the constraints on the particle's behavior or, dually, on the freedom that the setup allows the particle. Each relation will lead to a poset, which is dual to the poset ordered by its dual ordering relation. Maintaining a positive attitude, I will use the ordering relation that describes the freedom that the setup allows, so that setups with greater freedom will include setups that allow only a portion of this freedom. Thus at the top of this poset (Figure 4) is the setup $\top$ that has no obstructions placed between x_i and x_f. The particle having been prepared at x_i and detected at x_f is free to have taken any path during its traversal. Caticha uses a concise notation to describe setups. The top setup can be written as $\top \equiv [x_f, x_i]$, which reading right to left states that the particle is known to start at x_i and is known to arrive at x_f.

The most constrained setup is the one at the bottom, $\perp$, as it is completely filled with obstructions.[8] This setup is analogous to the logical absurdity in the sense that there is no way for a particle prepared at x_i to be detected at x_f. More interesting are the join-

---

[7] These paths are in space-time. For simplicity, I will often omit reference to the time coordinate.

[8] All setups with obstructions that provide no way for a particle to travel from x_i to x_f belong to the same equivalence class of obstructed setups and are represented by $\perp$. This includes setups with obstructed pathways.

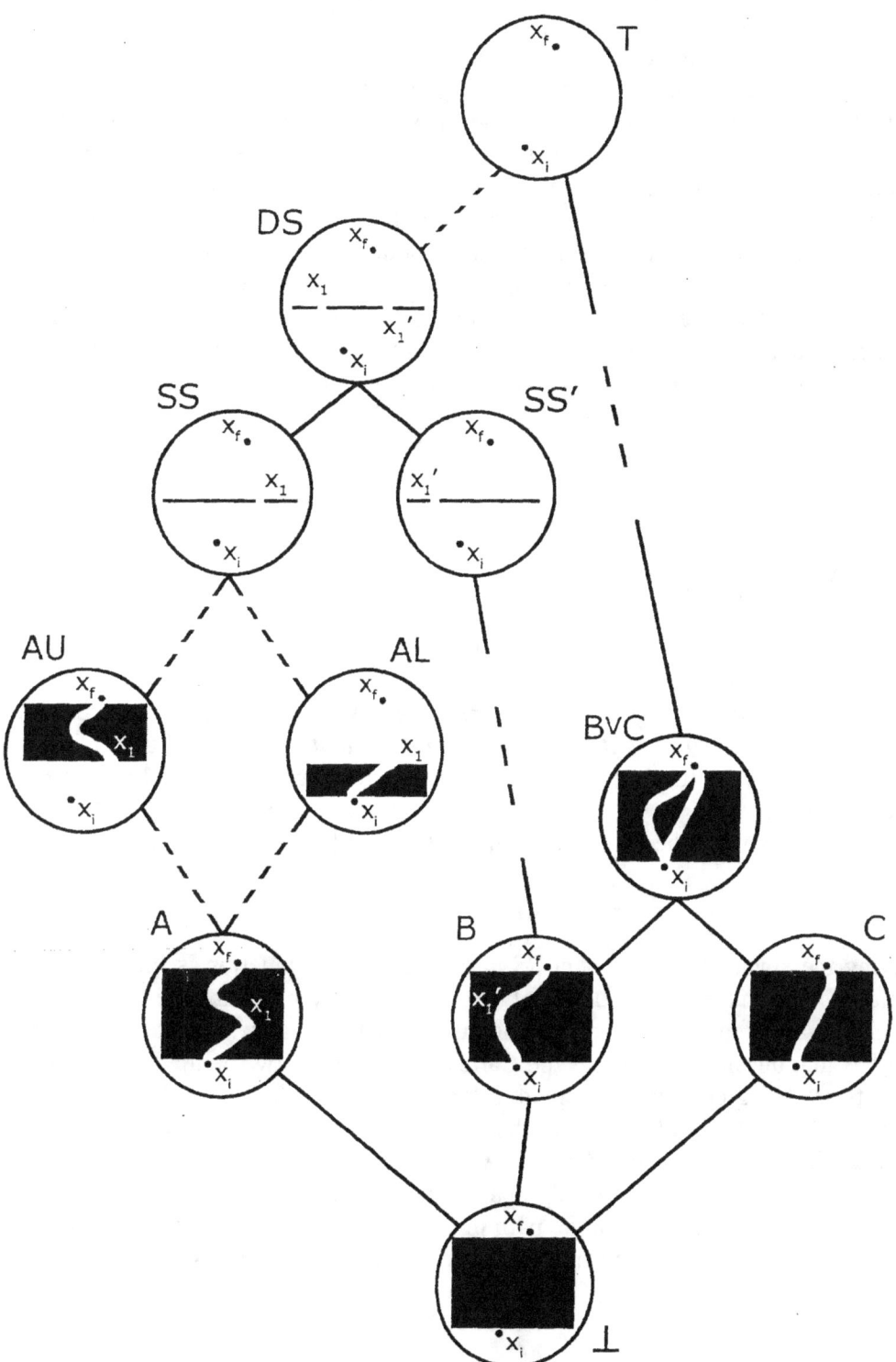

FIGURE 4. A partial schematic of Caticha's poset of experimental setups where a particle is prepared at x_i and is later detected at x_f. Solid lines indicate $\prec$, whereas dashed lines indicate $\leq$ where there are setups that are not illustrated. See the text for details.

irreducible setups that cover $\perp$. These are the setups that allow only a single path from x_i to x_f. Three examples A, B, and C can be seen in Figure 4.

All the setups in the poset can be generated by joining the join-irreducible setups. As Caticha, defined his poset in terms of an algebra, we must look at this algebra carefully to identify the complete set of join-irreducible setups. Obviously the atomic setups, which are the setups with a single path from x_i to x_f, are join-irreducible. However, Caticha defined the join of two setups only for cases where the two setups differ by at most one point at one time. Thus we must work through his algebra using the poset structure to discover how to perform the join in more general cases. Consider the two setups SS and SS' in Figure 4. The setup SS is a single-slit experiment where the particle is known to have been prepared at x_i, is known to have gone through the slit at x_1 and was detected at x_f, written concisely as $SS = [x_f, x_1, x_i]$. Similarly, SS' is a different single-slit setup written $SS' = [x_f, x_1', x_i]$. Their join is a double-slit setup

$$DS = SS \vee SS' \tag{96}$$

found by

$$DS = [x_f, x_1, x_i] \vee [x_f, x_1', x_i], \tag{97}$$

which is written concisely as

$$DS = [x_f, (x_1, x_1'), x_i]. \tag{98}$$

This can be read as: '*The particle was prepared at x_i, is known to have passed through either the slit at x_1 or the slit at x_1', and was detected at x_f*'. This definition describes the join of two setups differing at only one point. Before generalizing the join, we will first examine the meet.

Caticha describes the simplest instance of a meet [11, Eqn. 2], which is

$$[x_f, x_1] \wedge [x_1, x_i] = [x_f, x_1, x_i], \tag{99}$$

giving the single-slit experiment SS in Figure 4. This equation is interesting, because neither setup on the left-hand side of (99) is a valid setup in our poset, since the particle is not always prepared at x_i and it is not always detected at x_f. Instead, what is going on here is that this meet represents the Cartesian product of two setups. Two smaller setup posets are being combined to create a larger setup poset. This is demonstrated by the meet of AU and AL in Figure 4, each of which has a single path on one side of x_1 and is free of obstructions on the other side. AU is the Cartesian product of the top setup in the set of posets describing particles prepared at x_i and detected at x_1, which I write as $\top_{1i}$, and a setup consisting of a single path in the set of posets describing particles prepared at x_1 and detected at x_f. We can write these posets coordinate-wise

$$\begin{aligned} AL &\equiv (\top_{f1}, A_{1i}) \\ AU &\equiv (A_{f1}, \top_{1i}) \end{aligned} \tag{100}$$

where A_{1i} is the setup where the particle is prepared at x_i and is detected at x_1 and is constrained to follow that portion of the path in the single-path setup A. Setups A_{f1}

and $\top_{f1}$ are defined similarly. Their meet is found trivially, since for each coordinate, $x \wedge \top = x$ for all x. Thus

$$AU \wedge AL = (A_{f1}, A_{1i}) \tag{101}$$

which is a valid expression for the entire single-path setup A

$$A = (A_{f1}, A_{1i}), \tag{102}$$

and is consistent with Caticha's definition in (99). This interpretation, which should be compared to our earlier discussion on the lattice product, also provides insight into how one can decompose the single-path setups A, B, and C into a series of infinitesimal displacements. Each infinitesimal displacement of a particle can be described using a setup in a smaller poset. Using the Cartesian product, these setups can be combined to form larger and larger displacements. Thus the single-path setups can be seen to represent meets of many shorter single-path segment setups.

We can use what we have learned from the product space example to better understand the join of two experimental setups. We can join the setups AL and AU using the Cartesian product notation. Since $\top$ is the top element of the poset, when joined with any other element of the poset the result is always the top, i.e. $\top \vee x = \top$, for any x in the poset. Thus we see that

$$
\begin{aligned}
AU \vee AL &= (A_{f1}, \top_{1i}) \vee (\top_{f1}, A_{1i}) \\
&= (A_{f1} \vee \top_{f1}, \top_{1i} \vee A_{1i}) \\
&= (\top_{f1}, \top_{1i}) \\
&\equiv SS.
\end{aligned}
\tag{103}
$$

The interpretation of the two posets forming this Cartesian product is key to better understanding the join. The result states that the particle is prepared at x_i and is free to travel however it likes to be detected at x_1, and that it is prepared at x_1 and is free to travel however it likes resulting in a detection at x_f. The result is a particle that is prepared at x_i, passes through x_1 and is detected at x_f. Thus their join is the single slit experiment, $AU \vee AL = SS$. This result extends Caticha's algebraic definition of the join.

Two setups with non-intersecting paths can also be joined. This must be the case since the top setup in this poset is obstacle-free and includes all single-path setups. Consider a setup divided by the straight line connecting x_i and x_f. Two setups can be formed from this, one by filling the left half with an obstacle, call it L, and the other by filling the right half R. Clearly, their join must be the top element, as one setup prevents the particle from being to the left of the line and the other one prevents it from being to the right. This can be extended by considering two setups each with paths that do not intersect one another. This example is shown in Figure 4, as the join of two single-path setups B and C, which results in a two-path setup $B \vee C$.

At this stage, it is not completely clear to me how to handle setups with two paths that intersect at multiple points. Clearly, two paths that intersect at a single median point, such as x_1, can be considered to be the product of two setups each with two non-intersecting paths. Again, consistency will be the guiding principle. What is important to this present development is that we know enough of the algebra to see what kind of laws we can derive from order theory by generalizing from order-theoretic inclusion to degrees of inclusion. First, Caticha showed that the join is associative. This implies that there exists a sum rule. Second, he showed that when the setups exist the meet is associative and distributes over the join. This leads to a product rule. By measuring degrees of inclusion among experimental setups with complex numbers, Caticha showed that the sum and product rules applied to complex-valued valuations are consistent with the rules used to manipulate quantum amplitudes. By looking at the join of setups B and C one can visualize how application of the sum rule leads to Feynman path integrals [36], which can be used to compute amplitudes for setups like $B \vee C$, and by iterating across the poset, SS and T. Also, the amplitude corresponding to a setup representing any finite displacement can be decomposed into a product of amplitudes for setups representing smaller successive displacements. Last, it should be noted that the lack of commutativity of the meet implies that there is no Bayes' Theorem analog for quantum amplitudes.

Furthermore, using these rules Caticha showed that that the only consistent way to describe time evolution is by a linear Schrödinger equation. This is a truly remarkable result, as no information about the physics of the particle was used in this derivation of the setup calculus—only order theory! The physics comes in when one assigns values to the setups in the poset. This is done by assigning values to the infinitesimal setups, which is equivalent to assigning the priors in probability theory. At the stage of assigning amplitudes, we can now only rely on symmetry, constraints, and consistency with other aspects of the problem. The calculus of setup amplitudes will handle the rest. The fact that these assignments rely on the Hamiltonian and that they are complex-valued are now the key issues. Looking more closely at the particular symmetries, constraints and consistencies that result in the traditional assignments independent of the setup calculus will provide deeper insight into quantum mechanics and will teach us much about how to extend this new methodology into other areas of science and mathematics.

DISCUSSION

Probability theory is often summarized by the statement '*probability is a sigma algebra*', which is a concise description of the mathematical properties of probability. However, descriptions like this can limit the way in which one thinks about probability in much the same way that the statement '*gravity is a vector field*' limits the way one thinks about gravitation. To gain new understanding in an area of study, the foundations need to be re-explored. Richard Cox's investigations into the role of consistency of probability theory with Boolean algebra were a crucial step in this new exploration. While Cox's technique has been celebrated in several circles within the area of probability theory [33, 23, 37], the deeper connections to order theory discussed here have not yet been fully recognized. The exception is the area of geometric probability where Gian-Carlo Rota has championed the importance of valuations on posets giving rise to an area of mathematics which ties together number theory, combinatorics, and geometry. Simple relations, such as the inclusion-exclusion principle, act as beacons signalling that order theory lies not far beneath.

Order theory dictates the way in which we can extend an algebra to a calculus by assigning numerical values to pairs of elements of a poset to describe the degree to which one element includes another. Consistency here is the guiding principle. The sum rule derives directly from consistency with associativity of the join operation in the algebra. Whereas, the product rule derives from consistency with associativity of the meet, and consistency with distributivity of the meet over the join.

It is clear that the basic methodology of extending an algebra to a calculus, which is presently explicitly utilized in probability theory and geometric probability, is implicitly utilized in information theory, statistical mechanics, and quantum mechanics. The order-theoretic foundations suggest that this methodology might be used to extend any class of problems where partial orderings (or rankings) can be imposed to a full-blown calculus. One new example explored at this meeting is the *ranking of preferences* in decision theory, which is explored in Ali Abbas' contribution to this volume [38]. More obvious is the relevance of this methodology to the development of the calculus of inquiry [25, 26, 8, 6], as well as Bob Fry's extension of this calculus to cybernetic control [27]. A serious study of the relationship between order theory and geometric algebra, recognized and noted by David Hestenes [39, 40], is certain to yield important new results. With the aid of geometric algebra, an examination of projective geometry in this order-theoretic context may provide new insights into Carlos Rodríguez's observation that the cross-ratio of projective geometry acts like Bayes' Theorem [5].

ACKNOWLEDGMENTS

This work supported by the NASA IDU/IS/CICT Program and the NASA Aerospace Technology Enterprise. I am deeply indebted to Ariel Caticha, Bob Fry, Carlos Rodríguez, Janos Aczél, Ray Smith, Myron Tribus, David Hestenes, Larry Bretthorst, Jeffrey Jewell, and Bernd Fischer for insightful and inspiring discussions, and many invaluable remarks and comments.

REFERENCES

1. Boole, G., *Dublin Mathematical Journal*, **3**, 183–198 (1848).
2. Boole, G., *An Investigation of the Laws of Thought*, Macmillan, London, 1854.
3. Cox, R. T., *Am. J. Physics*, **14**, 1–13 (1946).
4. Cox, R. T., *The Algebra of Probable Inference*, Johns Hopkins Press, Baltimore, 1961.
5. Rodríguez, C. C., "From Euclid to entropy," in *Maximum Entropy and Bayesian Methods, Laramie Wyoming 1990*, edited by W. T. Grandy and L. H. Schick, Kluwer, Dordrecht, 1991, pp. 343–348.
6. Knuth, K. H., *Phil. Trans. Roy. Soc. Lond. A*, **361**, 2859–2873 (2003).
7. Graham, R. L., Knuth, D. E., and Patashnik, O., *Concrete Mathematics, 2nd ed.*, Addison-Wesley, Reading, Massachusetts, 1994.
8. Knuth, K. H., "What is a question?," in *Bayesian Inference and Maximum Entropy Methods in Science and Engineering, Moscow ID, USA, 2002*, edited by C. Williams, AIP Conference Proceedings 659, American Institute of Physics, New York, 2003, pp. 227–242.
9. Davey, B. A., and Priestley, H. A., *Introduction to Lattices and Order*, Cambridge Univ. Press, Cambridge, 2002.
10. Birkhoff, *Lattice Theory*, American Mathematical Society, Providence, 1967.
11. Caticha, A., *Phys. Rev. A*, **57**, 1572–1582 (1998).
12. Aczél, J., *Lectures on Functional Equations and Their Applications*, Academic Press, New York, 1966.

13. Rota, G.-C., *Studies in Pure Mathematics*, (Presented to Richard Rado), Academic Press, London, 1971, chap. On the combinatorics of the Euler characteristic, pp. 221–233.
14. Rota, G.-C., *The Mathematical Intelligencer*, **20**, 11–16 (1998).
15. Klain, D. A., and Rota, G.-C., *Introduction to Geometric Probability*, Cambridge Univ. Press, Cambridge, 1997, ISBN 0-521-59654-8.
16. Rota, G.-C., *Zeitschrift für Wahrscheinlichkeitstheorie und Verwandte Gebiete*, **2**, 340–368 (1964).
17. Krishnamurthy, V., *Combinatorics: Theory and Application*, John Wiley & Sons, New York, 1986, ISBN 0-470-20345-5.
18. Barnabei, M., and Pezzoli, E., *Gian-Carlo Rota on combinatorics*, Birkhauser, Boston, 1995, chap. Möbius functions, pp. 83–104.
19. Möbius, A. F., *J. reine agnew. Math.*, **9**, 105–123 (1832).
20. McGill, W. J., *IEEE Trans. Info. Theory*, **4**, 93–111 (1955).
21. Bell, A. J., "The co-information lattice," in *Proceedings of the Fifth International Workshop on Independent Component Analysis and Blind Signal Separation: ICA 2003*, edited by S. Amari, A. Cichocki, S. Makino, and N. Murata, 2003.
22. Pölya, G., and Szegö, G., *Aufgaben und Lehrsätze aus der Analysis, 2 vols.*, Springer, Berlin, 1964, 3rd edn.
23. Tribus, M., *Rational Descriptions, Decisions and Designs*, Pergamon Press, New York, 1969.
24. Smith, C. R., and Erickson, G. J., "Probability theory and the associativity equation," in *Maximum Entropy and Bayesian Methods, Dartmouth USA, 1989*, edited by P. F. Fougère, Kluwer, Dordrecht, 1990, pp. 17–30.
25. Cox, R. T., "Of inference and inquiry, an essay in inductive logic," in *The Maximum Entropy Formalism*, edited by R. D. Levine and M. Tribus, The MIT Press, Cambridge, 1979, pp. 119–167.
26. Fry, R. L., *Maximum entropy and Bayesian methods*, Johns Hopkins University (1999), electronic Course Notes (525.475).
27. Fry, R. L., "Cybernetic systems based on inductive logic," in *Bayesian Inference and Maximum Entropy Methods in Science and Engineering, Gif-sur-Yvette, France 2000*, edited by A. Mohammad-Djafari, AIP Conference Proceedings 568, American Institute of Physics, New York, 2001, pp. 106–119.
28. Knuth, K. H. (2004), in preparation. The calculus of inquiry.
29. Jaynes, E. T., *IEEE Trans. Syst. Sci. Cyb.*, **SSC-4**, 227 (1968).
30. Aczél, J., Forte, B., and Ng, C. T., *Adv. Appl. Prob.*, **6**, 131–146 (1974).
31. Hartley, R. V., *Bell System Tech. J.*, **7**, 535–563 (1928).
32. Shannon, C. F., and Weaver, W., *The Mathematical Theory of Communication*, Univ. of Illinois Press, Chicago, 1949.
33. Jaynes, E. T., *Physical Review*, **106**, 620–630 (1957).
34. Tribus, M., *Thermostatics and Thermodynamics*, D. Van Nostrand Co., Inc., Princeton, 1961.
35. Jaynes, E. T., "Where do we stand on maximum entropy," in *The Maximum Entropy Formalism*, edited by R. D. Levine and M. Tribus, The MIT Press, Cambridge, 1979, pp. 15–118.
36. Feynman, R. P., *Rev. Mod. Phys.*, **20**, 367–387 (1948).
37. Jaynes, E. T., *Probability Theory: The Logic of Science*, Cambridge univ. Press, Cambridge, 2003.
38. Abbas, A. E., "An information theory for preferences," in *Bayesian Inference and Maximum Entropy Methods in Science and Engineering, Jackson Hole WY, USA 2003*, edited by G. J. Erickson, American Institute of Physics, New York, 2003.
39. Hestenes, D., and Ziegler, R., *Acta Appl. Math.*, **23**, 25–63 (1991).
40. Hestenes, D., *Acta Appl. Math.*, **23**, 65–93 (1991).